ALPHA AND OMEGA

A Science Fantasy

By

Dorothy L. Hernandez

Illustrations by

Mary Redman

ISBN (e-Book): 979-8-2954-7779-9

ISBN (PB): 979-8-2954-7778-2

ISBN(HB): 979-8-2954-7777-5

Dorothy Hernandez was born and raised in the Southern part of Utah, Arizona and Nevada.

As a child she would wander off into the hills picking wild flowers. Dorothy had an imaginative childhood and sparkles as she talks. It is the same sparkle as an adult allowed into the realm of childhood.

Mrs. Hernandez's first love, however, is children, of whom she says: "They are the tender young buds that blossom in the night...

and if I have touched but one little heart, then I will have made life worthwhile."

To Contact Dorothy

Dorothy L. Hernandez

E-mail - dnp@teknett.com

Website - www.teknett.com/pwp/dnp

Preface

In a galaxy beyond the Milky Way is a planet called "Plateo." Ravished by a nuclear war, the people of the Federated Zones went underground until they could rebuild their planet.

Great losses were suffered, including the extinction of the only living dinosaurs known to any galaxy. The leader of the Federated Zones, President Karabe, sent his daughter and a crew of men to Dinosaur Island, where they hunt for dinosaur eggs that have not been contaminated or destroyed. After a harrowing experience they find two undamaged eggs. One is a male and one is a female. President Karabe names them Alpha and Omega, which when translated means "the beginning and the end."

Since Plateo is similar to Earth, they have decided to deposit the eggs in a cave in a remote area of the Rocky Mountains. In the meantime, a place will be prepared for the dinosaurs on Plateo.

After a probe is sent out, the Federated Zones discover that their enemies, the Paradons, have not been destroyed. In fact, they plan to invade the Earth. A battle is waged in space.

Once victory is achieved, a well-deserved holiday to the planet Gaeleo proves to be fun and relaxing.

Early in the springtime the eggs hatch. Two small, lovable dinosaurs emerge.

In late August a Boy Scout troop from Glennview, about five miles away, are having their annual camping trip. Two young boys, Skeeter and Doby, and their guide Henry, discover the baby dinosaurs. However, their story is so remarkable that it is not believed.

After the second summer passes, the boys go back to see the dinosaurs. Now quite large, the dinosaurs follow them home. Even though the boys try to hide from the great beasts, there is no escape.

Walking through barns and fences and water towers, the dinosaurs threaten to inadvertently destroy the farming community of Glennview. Plateo comes to their rescue. Many moments of fear, then triumph, along with joy, conclude this wonderful heartwarming story.

Chapter 1

In a galaxy beyond the Milky Way is a planet called "Plateo." Similar to our Earth, this planet had continents with rivers and oceans, mountains and deserts. Vegetation grew abundantly. Life flourished. Plateo had everything necessary to sustain the balance of nature. Even the people are much the same as we are. However, they are more advanced in medicine and space technology.

Despite their intelligence, a never ending struggle for power existed among the countries of Plateo. Not unlike the planet Earth, the threat of a nuclear war always existed.

Suddenly and without warning the dreaded event happened. A blinding flash of light made the sun look pale as it burned and seared across the land. This was only the beginning. An earsplitting explosion rocked the very foundations of Plateo. Buildings and mountains crumbled. Giant tidal waves rose without mercy and beat against the helpless shores. Chilling screams filled the air as people ran defenseless toward underground shelters. Devastation and loss was everywhere. Plateo became silent. In that silence, a large mushroom cloud rose up in defiance and darkened the entire sky. It was soon followed by a grayish-white ash which fell like deadly snowflakes and covered the planet in a blanket of death.

Deep in a cavern of an underground shelter was the President of the Federated Zones. His name was Karabe. He called for a meeting of his people. The first words he spoke were:

"And now my people, may we live in peace."

Everyone listened with interest, for they loved the small, slight man, with the long white beard that looked like corded wool. Whisks of white hair shaded the pink crown of his practically bald head. No one knew how old President Karabe was, for people lived to be very old on Plateo.

President Karabe continued on with his speech:

"Uppermost in all of your minds is the survival of our people. We did not suffer as great a loss as we once calculated. Our sick and injured are recovering. You may visit the computer room for data on loved ones in other zones."

The President was dressed as usual in his long white robes. They symbolized purity, love, devotion and loyalty. Tapping loudly with his cane and drinking a sip of water, the small man went on with his speech, his voice rising a notch.

"Will we survive? As it looks now, the answer is yes. However, this will take a lot of teamwork on everyone's part. We have food to last for many years. Our greenhouses are already in production. Water is our main concern. Every drop must be conserved and recycled."

People thronged the underground arena, hanging on to every word the President said.

"And now, my people, this is not the end, but a new beginning. Everyone will have a job to do. Believe me, there is much to be done. We have a whole new world to build.

"Outside of our underground shelter is death and destruction. Vultures are flying side by side with other animals and birds. Leaves no longer flutter on the thistle that was trees. Bees do not buzz. Ants do not crawl. Even the cockroaches have disappeared."

At this point, everyone laughed.

"Rainfall will help," the President said, "but no one can cleanse our planet, except us. Some of you will be sent into the mountains, others to the beaches, still others into the cities. You will take the ray guns and

disintegrate everything that is dead. When Plateo is ready, we will import birds and animals and even insects from other planets. Does everyone support me in this endeavor?"

The crowd clapped and cheered.

President Karabe wiped the perspiration from his brow. "As yet we have not heard from our enemies the Paradons. Their silence leads me to believe they have been defeated. We will launch our computer tracking probe and gather any and all information possible."

Speaking with great emotion, President Karabe's voice became shaky and his blue eyes filled with tears.

"One of my greatest sorrows is the extinction of our Plateosauruses. We had the only living dinosaurs known to any planet. I intend to send a special detail to Dinosaur Island, headed by my daughter Reva. This detail will clean up all the remains. Geiger counters will be used to search caves for eggs that have not been contaminated. In the event we are successful, the eggs will be flown to another planet for incubation. With our computer scanning devices we have searched the solar systems for an ideal planet. Our choice is the planet Earth. Since they have excavated bones and reconstructed

the skeletons of dinosaurs which lived millions of years ago, I feel they have a great appreciation of the animals. The people on Earth must not know of this project, however, because the eggs and the dinosaurs when they are hatched, could fall into the wrong hands. I have chosen a remote site in a secluded range of what is known as the Rocky Mountains.

We do not have to worry about detection. Our space ships have circled the Earth many times. Earthlings call our sightings 'UFO's' which stands for 'Unidentified Flying Objects.' So far, none of the governments or their militaries have backed anyone in their beliefs and/or reports of these sightings. Therefore, we are safe."

President Karabe ended his speech with this final message: "Check the work schedules. We will start on our projects tomorrow morning. Thank you, that is all."

Reva arose bright and early. Young and adventurous, her father had let her go to Dinosaur Island only twice in her lifetime. However, the lovely brunette with the big brown eyes, had become overwhelmed by the dinosaurs. Sometimes, she could think of nothing else.

Even though the dinosaurs were not dangerous, as they had no predators and were vegetarians, her father had tried to shield her from all danger. After all, they were big and clumsy. Some of them grew as high as 15 feet (4.5 meters) or more with a length of approximately 26 feet (7.8 meters) or more, including the tail. A person could get squashed to death under one foot, or swished unmercifully by their giant tails.

The triulon was quickly made ready. A triulon is similar to Earth's helicopter, only it can do much more. The highly developed craft can not only fly, but it can also maneuver on land and go under water for deep sea exploration. Everything was ready. Special capsules were made to carry the eggs in the cargo craft. Silver colored contamination suits and gas masks had been issued to everyone. A crew of ten men and Reva started out for Dinosaur Island. Large doors opened up from underground, and off they flew on their strange and unusual mission.

After about a forty minute flight, the triulon circled the island. Reva could see the huge grayish brown carcasses of the dinosaurs. Flying low, the men were able to disintegrate the bodies with their ray guns. Reva felt sad as one by one the giant lizards were destroyed. She could almost hear their bellowing voices. Reva noticed too, the damage that had taken place. Once green grasses were now burned and brown. Once fresh water streams ran murky with no course to follow, except to empty into dead pools of yellow and greenish looking slime. Trees were bare, like dead wood sticking out of the ground. Almost sorry she came, Reva was determined to see this mission accomplished.

Everyone put on their gas masks as the triulon landed in front of a large cave. The gray capsules were unloaded at the entrance. Carefully, the crew made their way inside. Not even a bat fluttered overhead, thousands however, lay dead on the floor. Naturally, they all had to be disintegrated before exploration could begin.

No signs of life emerged as their bright lights scanned the cave. Deeper and deeper they went. Suddenly the crew could see light filtering through the roof. Rocks and debris spread before them like stumbling blocks. Reva looked at the jagged walls. Running her gloved hand over the surface, she could see imprints of bugs and spiders etched forever in stone by the nuclear explosion. Now she knew, even if they found eggs, there would be no hope.

Searching further, they came to an area with a sandy floor. Sticking part way out of the ground, they spotted what appeared to be dinosaur eggs. Instead of having grayish-white shells, they were a tarnished brown and rust color. Some were even cracked showing the embryo inside. Reva felt she was going to be sick. She motioned to the men to destroy the ruined eggs. With heavy hearts, Reva and the crew made their way out of the cave.

Once aboard the triulon, everyone took off their gas masks and head gear. Reva shook off her long brown hair, which looked like silk floating on her shoulders. Her small, heart-shaped face looked puzzled and sad.

"What do we do now Ammone?" she asked.

Ammone was the head pilot. An expert in his field, he had been a space navigator and commanded a fighting squadron. His blue-gray eyes were deep in concentration. Rubbing his short, blond, wavy hair, he replied, "We still have two more caves to explore. I'd say, Reva, we should not leave the island until all hope is gone. You know how much this project means to your father."

Other crew members were grumbling and voicing their opinions. "We should go home right now."

"This mission is unreasonable. People should come before dinosaurs."

"We're out here subject to radiation and who knows what, just to satisfy the whims of Karabe."

Hearing all of this, Reva became furious. "We will go on Ammone."

"You heard the lady," Ammone said, as he stood angrily looking at the crew. "No matter what you men think, we are making the decisions."

The triulon once again took to the sky. Reva could see the huge boulders and rocks that had fallen from the mountains.

Dinosaur Island had been hit hard. Flying low once more, Ammone landed the Triulon in front of what once had been a cave. A dinosaur was lying there, half covered with rocks, as if his last effort was to survive.

"Here's one we missed, Reva," Ammone remarked. "I'll have the men destroy the carcass, but there's no way we can explore this cave. It has completely collapsed."

"Yes, I know," sighed Reva. "We have one more cave to go."

With all hope gone, the crew continued on. At the far end of the island, they could see the entrance to the third cave. A few rocks had fallen, but not enough to keep them from entering.

"This looks like a big cave," said Reva, addressing the crew. "Bring the capsules with you. We'll carry them as we go."

Once inside, Reva was astounded, for she had never seen such a mammoth cave. She strained her neck just to glimpse the rocky, creviced ceiling. Nature had carved out a mountain. Here and there were small pools of water. Dead bugs and spiders floated lifelessly upon them. Days and weeks could be spent here, thought Reva, before they found any eggs.

Even though the day was almost gone, the crew went on and on. Their bright lights kept searching and searching. Unexpectedly they spotted something moving beneath their feet. Stunned at finding anything alive, they all gasped. Ammone turned on his transmitter.

"This is a Maconda snake. Evidently, he has been hibernating under some rocks. See how pale his colors are. These snakes are quite harmless. However, this particular species feeds on insects and mice. To eat anything now, the snake would surely die. Our best bet is to destroy the Maconda and save him from a long and agonizing death."

At this time, Reva turned on her transmitter.

"Let's run the Geiger counter over him first, Ammone, and see if the snake is contaminated."

The men held the snake. To everyone's amazement, there nothing registered on the Geiger counter.

"We must save him at all costs," declared Reva. "Even though he is only a snake, this will be a first toward saving our world. There are contamination bags in the triulon. We brought them, remember, to gather rock samples for levels of radiation."

"Oh, yes, Ammone said, "good thinking."

When the men had returned with the bags, the penetration of the cave continued. Reva's spirits were high once more. She desperately wanted to find some eggs. Walking almost a quarter of a mile (quarter of a kilometer) further into the cave, sandy soil once more appeared. Excited, the crew searched frantically. Reva thought she spotted something. Falling to her knees, she began half brushing and half digging in the sand. Yes, there were eggs, two of them. Reva turned on her transmitter.

"Bring the spades, but make sure you do not damage the shells."

Slowly and gently they rolled the grayish-white eggs over. No one was taking any chances. Actually, the eggs were durable like a turtle's, hard shelled, thick and yet elastic. However, dinosaur eggs are much larger than a turtle's, some as long as 12 inches (30 centimeters) and 10 inches (25 centimeters) across with a 7-pint (3.3 liter) liquid capacity.

Reva held her breath as the Geiger counter measured for contamination. No, they didn't appear to be contaminated. Overjoyed, she could hardly wait to tell her father.

The men opened the capsules and deposited the eggs. Later on, they would be pressurized to create a vacuum and keep the eggs suspended within the capsule on their long journey to Earth.

Chapter 2

Darkness shrouded the planet as Reva and her crew returned home. A gentle, blessed rain was falling, a first since the war. President Karabe was pacing the floor in his green dinner jacket. Even though Ammone had transmitted ahead, he was anxious to see his daughter.

Reva hurriedly discarded her contamination suit and still in her flight jumpers, rushed to her father's quarters.

"Oh, Father!" cried Reva embracing him. Half crying and half laughing, her words came tumbling out. "Oh, Father! It's raining outside."

"Yes, I know, my Princess," he soothed as he stroked her hair and kissed her damp cheeks, which were already wet with tears. Happiness had never felt so good.

During the course of dinner, Reva related the events of the day to her father and mother. Reva's mother never spoke much, but love and compassion showed in her soft, brown eyes. Just a nod or a gesture could mean more than a room full of words. Always neat and articulate, her short, brown hair was laced with silver and waved lovingly around a small, charming face. Small, but not delicate, she was a pillar of strength for her husband and her daughter.

After relating the day's events, Reva excused herself and went to her room to retire. Feeling warm and snug and happy, she drifted off to sleep.

* * * * *

Reva was awakened by the sound system in her room. "This is a code three message. Report to the laboratory immediately." The message was repeated several times.

Hurriedly, she slipped into her lavender mini dress, wrestled into her silver boots, and brushed her hair. To call in this manner, Reva thought, it must be very important. A code three was almost a panic button.

"Yes, Father," gasped Reva, out of breath, "I'm here."

Her father was standing by the testing chambers. He had on a white jacket like a doctor and was writing in a table.

"Sit down at the table, Dear, and I will tell you

about our findings. We have just tested the Maconda

snake," said the President. "You will be happy to know,

Reva, that the snake is about to lay eggs."

"Oh, Father, this is more than we had hoped for." Reva was overjoyed.

"Her best news is yet to come." Her father's eyes sparkled as he spoke. "Our dinosaur eggs tested out perfect. One is a male and the other is a female. I have named the male Alpha, which means the beginning. I have named the female Omega, which means the end. Since they were here before man, I pray they will survive, even after we are gone. The eggs have been marked with the names I have chosen, first in English and then in Plateonian."

"This truly is a new beginning," Reva said. "When do I leave for the planet Earth?"

"It is imperative that you go within the week. We must get the eggs to the new planet before springtime comes to the Rockies. The eggs must hatch in their new environment. I know this is short notice, Princess, but time is of the essence."

"Excited, Reva went to her quarters to make ready for the long trip to Earth.

Her happiness was to be short lived however, as

Reva was paged once again. "This is a code four. Report to Federated Conference Room."

A council of twelve representatives was sitting at a big, round table with her father. Reva knew something was terribly wrong when she saw the worried, troubled looks on their faces. Puzzled, Reva had never been asked to attend a meeting of state. After she was seated, her father began to speak.

"As you can see, I have called an emergency meeting. I have asked Ammone and Reva to join us. The tracking probe we have sent to the Paradoon Nation has returned. They have not been annihilated as their silence would indicate. However, they do not intend to attack us again. We pose no threat. Instead of building a new world as we are doing, they plan to invade the Earth. Their departure is scheduled in three days."

Reva watched her father, who was still in his laboratory clothes. He suddenly looked old and tired. Too much responsibility was devouring the man. Reva's heart ached. Only an hour ago he had been bubbling over with joy.

His voice full of emotion, the President continued. "Earth, as you know, is a beautiful planet. No other except Plateo, that we know of, can compare to this perfect world. We have been afraid they would destroy themselves, much as we have. My conscience will not let the Paradons invade the Earth. Earth cannot fight back for they are not far enough advanced in space warfare to defend themselves. Since we have planned to take the dinosaur eggs, in favor of sending our mother space ship, with fighters, to destroy the Paradons. Talking with them in the past has been a waste of time. A surprise attack will be necessary, to defeat forever, these tyrants. I have appointed Ammone to command the fighter squadron. He will personally pick the men to go on this mission. Even though my daughter, Reva will be disappointed, I would rather she remained here. Her safety is uppermost in my mind, as are all my people."

Reva couldn't believe what she was hearing. How could her father deny her this assignment?

One of the council of twelve raised his hand. "Yes," Mr. Zerack, Zone Six, you may speak," said President Karabe.

A large man took the floor. Gray hair brushed his temples like a paint brush and made the man look distinguished. A wide face and firm, brown eyes fit with his stature. Still dressed in blue coveralls, Mr. Zerack had been maintaining a greenhouse. His voice was loud and husky.

"I feel the deep concern you have for your daughter, but isn't this really a family matter? I know we have never allowed our women in combat.

Our men do not permit it. However, she will not be fighting. So the final decision should rest with you and Mrs. Karabe."

All the council clapped their hands and shouted, "Aye!"

"It seems that I have been outnumbered," said Karabe. "Even so, this will not alter our laws in the future. Now if you're all in favor of destroying the Paradons once and for all, we'll adjourn."

The answer was unanimous. Combat was inevitable.

Many anxious moments confronted Reva. She had never disobeyed her father. Going to Earth for the first time was an experience she didn't want to miss, especially since they were going to land and, not just circle the planet. Everything seemed to be a risk. Even the assignment to Dinosaur Island could have been costly as they were subject to high radiation.

Reva felt she had a good argument that would work in her favor. At dinner time she presented her side of the story. She knew her father would listen for he was always understanding. After all, he was a leader. Even though there was sadness written in his eyes, Reva knew this was going to be painful for both of them.

Reva got right to the point, as she seated herself.

"I know, Father, that you are against my trip to Earth and I can understand your feelings as I am your only daughter. But this journey means one of the greatest adventures of my lifetime. Please grant me your permission and your blessing."

President Karabe reached over and clasped his daughter's hand. "My Princess, you have no idea what this involves. If you are captured the Paradons are treacherous. You could be tortured beyond belief."

"Yes, Reva," her mother spoke, "we want only your safety."

"But Father," Reva pleaded, "I have complete confidence in Ammone, as you do. Even in your own words Father, we all have to make sacrifices."

"I know you are a brave girl, Reva," her father said, "and I had no idea this project meant so much to you." President Karabe smiled. "How could I refuse my little Princess anything?"

Reva jumped from her chair like a child and planted a big kiss on her father's forehead, then rushed to her mother and did the same. Never had dinner tasted so good.

"Now, Reva," her father said, as he finished his evening meal, "I have instructions in my desk which you must take with you."

Reva followed her father to his library. "We must be prepared at all times." Reva listened intently to her father. "In the event you have to shuttle to another planet for safety, the dinosaur eggs can be frozen and held in limbo for an indefinite time. However, avoid this at all costs. Something could go wrong, even though this is a safe procedure, we don't want anything, no matter how remote, to interfere with the incubation of the eggs."

"I understand," Reva answered.

President Karabe went on with his instructions. "When you find a suitable cave for the eggs, seal all entrances with an invisible curtain. You see, my Dear, many wild animals still make the mountains their home.

Cougars and coyotes could break the shells and feed on

the eggs. We have inserted a trigger inside the shells. When the dinosaurs are hatched, this trigger will go off

and remove the invisible curtain automatically. The dinosaurs can then look for food. And now, my Princess, I must at this time reveal the true facts of this journey."

Again sadness filled his tired eyes. "My plan was for the federation to make the decision concerning your mission to Earth. It did not happen. You see, I can't assign anyone to take your place. Failing to respond to my obligations, would serve as a bad example. How can I expect my people to face danger, if we cannot?"

"Oh," Reva gasped, "You should have told me Father."

"I could not make you go, Princess. Your determination and persistence was the final decision. I'm very happy and proud to have a daughter like you. Even though I still wish you weren't going, you have my blessings."

"Thank you, Father," Reva understood.

Chapter 3

Gigantic doors slowly opened on the surface of Plateo. A mother ship as large as a city emerged from underground. Rising like a balloon, it whirled majestically in a myriad of colors. Oranges, red and blues and purples mixed like whipped cream as the circular craft rose to embrace the heavens.

Everything had been prepared down to the last detail. Searching for the Paradons was their first objective. Scanners began combing the vast expanse of space. Departure was to coincide with the enemy. So far nothing. Could this be a hoax to throw them off?

Reva stayed on the bridge with Ammone. A strategic crew worked side by side with the commander. All this fascinated Reva. Never had she seen such precision. With just the touch of a button several hundred fighters would move like pawns on a chess board and jockey into position. Where was the enemy?

All of a sudden Ammone shouted, "We've got them! We've got them!"

Reva looked into the scanner. Down below them floated a large brown craft that looked like a river barge.

"Their fueling up," said Ammone. "See the tentacles reaching from their ship to the gassy planet."

Reva went in for a closer look. The gassy planet had been a source of fuel for many years. Long tubes were projected to siphon the gases into the space ships. On board it was then automatically refueled to fuel the larger crafts.

A massive explosion ripped through space as the tentacles of the Paradoon ship were torn out beneath them. Almost immediately, fire squadrons formed and surrounded both the crafts. They were now in a position to wage an all out war.

Not letting the Paradons have the first target, the Federated Zones started firing. Abruptly, the Paradoon armada was upon them. Explosion after explosion bounced off the mother ship of Plateo. Churning and rocking like an angry sea the entire gyro system was threatening to destabilize.

Fear gripped Reva as she sat frozen on the bridge. No longer was this game a fascination. She wanted only to leave, to escape. In the back of her mind she could hear Ammone giving out orders. These were commands to fight.

"Zero five, ninety degrees, zero ten, fifty degrees, zero eight, thirty degrees."

On and on he went until one by one the Paradoon fighters began to fall into the gassy planet.

"Direct fire! Direct fire!" As if Ammone was on an all time high, his adrenaline was flowing. "I'm going for their mother ship," he said.

Blast after blast bombarded the box-shaped enemy craft. Large holes were ripped and torn in the ship. Fires erupted everywhere, like intermittent flashes of the sun. Crippled and devastated the Paradons were defeated. Ammone stretched and relaxed. His arms were folded behind his head. A smug, contented look spread across his face. He looked over at Reva, who was sitting next to him. She had barely regained her composure.

"I've transmitted the news to your father," said Ammone. "He is very pleased, also very relieved. Except for the time-out to fight the war, we are pretty much on schedule."

After such a shattering experience, Reva couldn't help but admire Ammone's strength and confidence. Reva began to ask questions.

"Did you send a crew to board the Paradoon craft? Were there any survivors? Will they be our prisoners or will they return to Plateo?"

Ammone barely gave it a answer. "No, we haven't boarded the craft. There has been no surge in power since we bombarded them. I do not believe there are any survivors."

"What will happen to the ship?"

"It will drift aimlessly in space until the gravity of another planet pulls the craft, like a magnet into its atmosphere. Almost certainly, the ship will burn and disintegrate upon entry."

"Oh!" Reva jumped up as it a light went on in her head. "I can't forget the eggs. If something has happened to them, my father will never forgive me."

Reva rushed down long corridors, traveled in a tube ejector and passed through many storage areas until she reached the chamber where the capsules were stored. She checked the gauges of the huge drawers that held the dinosaur eggs. Temperatures, humidity, levels of stabilization and heartbeat were all necessary factors to determine their safety. Much to Reva's satisfaction, the turmoil of the ship during the war had not altered anything. Very relieved, Reva returned to the bridge. Everyone was cheerful and laughing and happy. Even

Commander Gareth was smiling. Reva had only experienced his serious side. A tall man, with dark hair and steel blue eyes, he ruled his ship and the battalion of space fighters under him with strict discipline. He looked at Reva from his high- backed chair on the bridge. In a stern but friendly voice Gareth asked, "How would you like to go on a short holiday?"

"That would be nice," said Reva, "but where?"

"In a few hours we will be approaching a planet called Gaeleo." The commander stood up and pointed out just where it was located on the space map. "Not much of a planet, I'm afraid. Only a small portion is inhabited."

"He-re and there Reva could still hear chuckles. She began to wonder what was so funny.

Commander Gareth explained, "Gaeleo is a planet for amusement only. A king by the name of Windwinkel is the ruler and he is a clown. His wife Patima is the fat lady. They do nothing but party, twenty-four hours a day. People from planets all over our galaxy visit Gaeleo. I have taken your father there many times to trade. You see, they live on the barter and trade system. Therefore, no one has to work. All they do is have fun. A carnival, a circus and gambling is going on at all times."

"You won't believe what you see and hear," added Ammone. "The animals actually talk."

"Oh, come on now, Ammone," Reva shook her head. "You've got to be kidding." She didn't believe him.

"Legend contends that a mad scientist was stranded with a lot of animals. He became lonesome for conversation. Out of desperation, he invented a serum which developed the glands around the larynx so the animals could speak. Just wait," said Ammone. "You'll see."

"Since my men have been in combat," said the Commander, "they deserve a little fun. We will shut down and give them some relaxation."

"This will not throw us off schedule?"

"No Reva," Gareth assured her. "You'll be on Earth by May."

True to form, Gaeleo, was indeed a very unique planet. Upon their arrival, King Windwinkel mounted his cushioned throne. He was dressed in a yellow clown suit with large purple buttons. His ears were floppy and he had a red flaming ball for a nose. Perched on his head was a pointed party hat. Windwinkel even laughed when he talked.

"Welcome, welcome to Gaeleo. Come into my tent," the funny man beckoned. "Meet my wife Patima."

Patima could spread across two thrones, she was so large. A whole bolt of cloth must have been used to make her shiny red dress. Short, curly brown hair hung in ringlets around her broad face. She sat and smiled.

Reva couldn't help but notice the contrast as she was with Gareth and Ammone. Both were dressed in their dark blue military uniforms and looked very handsome. Reva was in a pink, ruffled, mini dress. Not exactly a party outfit, but it would do.

"I've met you before," Windwinkel said to Gareth.

"Yes, I was here with President Karabe. This is his daughter, Reva."

"Oh yes." The little man laughed and shook her hand. "I'm happy to make your acquaintance."

"We have nothing to trade this time," he commented Gareth, "but we do have gold pieces."

"Everyone can use gold," the little man chuckled. "Now let me show you around."

Just then Reva spotted a beautiful Angora cat lying on a pillow. She bent down and gently stroked the animal's fur.

"You have nice hands," said the cat, in a high shrill voice.

Reva backed off. She was both startled and amazed.

"See what I told you?" teased Ammone, "You wouldn't believe me."

Bouncing along like a jumping jack, Windwinkel took them down streets filled with gaudy little trinket shops. People milled around everywhere. Some were dressed for a Mardi Gras, while others wore their native costumes. Finally, they arrived at the circus.

A man was trying to get a mule to perform.

"I'll get my bullwhip if you don't move."

In a low, gruff voice, the mule replied, "I'll kick you, if you do."

Now Reva knew what an animal would sound like if it could talk.

Up above their heads a peacock was perched on a wire. As she spread her rainbow colored plumage, she sang out, "I'm the prettiest bird alive. I'm the prettiest bird alive."

A band was playing at the circus. Animals were performing under a big orange tent. People clapped and cheered, as they did their tricks. Lions were sprawled out lazily in a fenced area. One lion yawned and growled, and in a loud, vibrant voice, he said, "I'm so hungry, I could eat a horse."

Just then a horse passed by. He looked down at the big cat and said, "Don't look at me."

All this was very amusing. Reva couldn't remember when she had enjoyed herself more. Upon conclusion, the monkeys did high wire acts.

"That's all folks," they said in unison.

Indeed that was all. Time had run out. The space ship was waiting.

A long, tedious journey lay ahead, as they traveled through the vastness of space. Reva could see the twinkling of planets, when the sun shined on them. Everyone found something to do. Some played games on electronic computers, while others watched pre-taped shows and read books. Reva had her job to do. She constantly checked the gauges of the dinosaur eggs.

Abruptly, half way to Earth, Reva was awakened out of a sound sleep. The mother ship began to rock and sway, as it did during the Paradoon war. Loud, pounding noises like a rain of pellets bombarded the craft. She quickly dressed. Despite the rolling of the ship, which made it difficult to walk, Reva made her way to the bridge.

Everyone was at their controls. Commander Gareth looked up from his chair as Reva entered. In a shaky and frightened voice she questioned the man. "Did the Paradons return? Are we being attacked?"

"Oh, no," Gareth replied. "We are experiencing a storm in space." He motioned for Reva to move closer. "Come and look for yourself."

Reva stood and looked into the monitor. Large boulders as big as a house seemed to be coming right at them. Gareth explained, "These are explosions from a sun. Every so often, these explosions occur. When they do, chunks of matter are hurled into space. They are called meteoroids. Sometimes they hit small comets and bounce them out of orbit. In turn, this causes them to break into pieces and they are flung in all directions. Eventually, the meteoroids are pulled into the gravity of other planets. Once there, they burn and disintegrate before entering an atmosphere."

"So far, the meteoroids have only caused us a little discomfort. However, our ship's guidance system is set to avoid hitting them. You have nothing to worry about." Gareth assured her.

In the days that followed, Reva looked at many galaxies and many planets.

Chapter 4

Excitement mounted as the mother ship approached the Earth. Reva sat breathless on the bridge. Her eyes were glued to the monitor. Even in her wildest imagination, Reva couldn't envision the spectacular beauty that was Earth. Continents were green and lush, lending a contrast to the frozen, white snowcaps that gripped the North and South poles. Blue, aquamarine oceans separated the land. White, fluffy clouds appeared, like fresh-popped corn, as they went along. No one but the Master, Reva thought, could have made such a perfect world.

Commander Gareth came over to sit with Reva. He began to speak. "No description that I, or anyone else, could convey to you, would describe what you are now seeing."

"Yes, Commander," Reva agreed, "You are right. No wonder the Paradons wanted to invade the Earth."

Gareth went to pour himself and Reva a cup of coffee. When he returned, he talked to Reva again. "Our sensor system has been jammed to all of Earth's detection equipment. Also, our computers know the altitude of all aircraft at all times. Our presence here will not be discovered. We will circle the Earth several times before taking the eggs down. You will get the grand tour."

In her soft, sweet voice Reva replied, "I would like that, very much."

Orbiting the Earth was quite an experience. Commander Gareth pointed out to Reva various places of interest as they went along.

"Now this continent is called South America. Look at the large clusters of green. Earthlings call that a jungle. See the line that meanders like a snake almost the length of the continent. That is called the Amazon River."

Moving farther north, Gareth proceeded to tell Reva about North America and the United States. Down below she could barely see the cities with their tall buildings and the massive web of freeways.

"What is the grayish-brown cloud," Reva asked, "that surrounds the cities?"

"Earth people call that pollution," replied Gareth. "They still have petroleum powered vehicles and unvented factories. Some of the water is polluted too."

"Oh what a shame," Reva gasped. "Aren't they aware of the damage they are doing to their beautiful planet?"

Commander Gareth continued, "See this long range of mountains? And he pointed them out. "Here is where you'll take the eggs."

Soon they were passing over the plains and moving east toward the Atlantic Ocean. Reva could see the great city of New York. Suddenly, a statue on an island caught her eye.

"What is that statue?" she asked Gareth. "I've never seen anything quite like it."

"I believe it is a symbol of freedom. Let me press a few buttons. We have all of Earth's history on our computers. Yes, here is the story. But wait, Reva, something else is showing on the computer. I believe this will explain what the statue is all about, without reading the entire history."

Gareth and Reva read the following poem.

A Lady Is Waiting

Alone and undaunted she's waiting,

On a beautiful glorious isle.

A beacon of light is held high in her hand,

On her face is a welcoming smile.

Alone she has gathered the lost sheep,

The oppressed, the homeless and strayed.

She's a vision to the war-weary soldier,

As she reigns o'er the land of the brave.

Alone she will guide us to freedom,

And the right to speak and to pray.

She will give the hope to the hopeless,

'Neath the flag of the great U.S.A.

Although she is only a statue,

And a gift from across the great sea.

She stands as a queen, so straight and serene,

O'er the land of the brave and the free.

So when you enter that wonderful harbor,

And your heart swells with heavenly mirth.

Fall down on your knees and thank your Creator,

That freedom is still on this Earth.

Reva and Gareth were both quite impressed.

By now Europe was looming in the distance. Gareth drew Reva's attention to the Rhine River in Germany and the Alps in Switzerland. Going further south and approaching Africa, the space craft lowered slightly to give Reva a better view of the giant pyramids in Egypt.

Darkness was crowding in on the African continent. They could barely make out the Himalayas with Mt. Everest, Earth's highest mountain range. On and on they traveled until the daylight appeared once more.

"In order to see the rest of the Earth," Gareth told Reva, "we must reverse our direction."

Gareth pushed a button and in the twinkling of an eye, the craft had turned. Circling and spinning like a top, they continued on their journey. Groups of islands greeted them as they began the tour of Asia. Just as before, Gareth lowered the space craft to give Reva a better view of interesting landmarks. She was fascinated by the city of Hong Kong and marveled at the Great Wall of China. Next was Australia. Surrounded by water, Australia seemed forlorn and isolated, except for a few islands. However, most of her beauty had been untouched by modern man.

Jungles and rivers and lakes filled the monitors as they spun over the lower part of Africa. Gradually going north, Reva was surprised at the dry Sahara Desert, which lent a contrast to the entire continent.

Commander Gareth leaned back in his chair. "Now Reva," he said, "you've had the grand tour. Get some rest and we'll take the eggs down."

Reva dressed in her warm jump suit and heavy shoes. She walked to the hangars where all the aircraft were stored. Ammone was there to greet her.

"Yes, Ammone, I'm excited. I can hardly wait to land on Earth."

Ammone began giving instructions. "Commander Gareth will hover the mother ship as close as possible. A red, flashing light will be our signal to go."

Off they flew over hills and valleys and pine trees. Giant mountains loomed before them, like walls of rock that reached the sky. Streaks of burnt sienna, umbers and ambers in broken veins throughout the grayish rock formations. Clouds were everywhere, on the mountain tops, like frosting on a cake.

At long last, Reva stepped on the living soil that was Earth. She breathed in deeply the cool, clean air. A soft breeze whispered in the trees. Birds sang. Lizards scampered here and there, startled at her arrival. A small stream rippled over stones and pebbles and ran toward a meadow. Clear, cold water, stomach and drank the pure, sweet water right out of the stream. Her expectations were fulfilled.

Ammone and the crew waited by the triulon for Reva. She brushed off her clothes and rejoined the men.

"What do you think of this idea?" Reva asked. "Half the men will look south and the other half will look to the north. We must find the right cave. Our plan calls for at least 26 feet (8 meters) high or more, with plenty of room for the dinosaurs, when they are grown, to move around. This will make the animals more secure, especially when they hibernate. Also, in the event there is flooding, they will not drown."

"Yes, that is a good idea," Ammone agreed. "I'll leave one man on guard at the triulon. We'll come back for the capsules."

Checking their individual flying gear before leaving, one by one they jumped into the air. Off they flew like strange birds gliding along. Small tunnels and arches could be found everywhere. Reva noticed a large slab of rock formation. The top was smooth and as large as a football field. She began to think. When they picked up the dinosaurs this would be an excellent place to transport them aboard the mother ship.

Looking around, Reva spotted a cave almost directly across from the rocky plateau.

She signaled Ammone on the transmitter.

"I believe I have found what we are looking for," Reva said.

"Good," Ammone replied, "I'll be right with you."

Reva landed in front of the cave. Soon Ammone and several crew members joined her. A large, spacious cavern with a small pool of water greeted their searching eyes. Yes, this was exactly what Reva had been looking for. One of the men fired the measuring gun. Registering over 30 feet (10.16 meters) high, they knew this would be the dinosaur's home.

Ammone flew the triulon close by the entrance. Unloading began. Seals on the capsules were broken and the eggs were placed on the sandy soil.

"According to my father's instructions," Reva told the crew, "we must put an invisible curtain over the entrance."

Everyone stepped outside. Ammone took a special laser beamer and sealed the cave. A sigh of relief could be heard among the men.

"We are almost finished," Reva said. "Our last instructions are to implant mushrooms and around the mountain. They are sensors that will pick up our beams from the space ship. When beamed upon, they will beep. In this way, we will know the exact location of the dinosaurs when the time has arrived to take them home."

Lasers again were used to penetrate the base of the mountain. Small, and looking like a mushroom, the sensors became embedded in the rock.

Reva was sorry they had to leave. Her only consolation was in the knowledge they would return. Rising once again, the mother ship spun and ascended into space.

* * * * *

Several weeks passed before any change took place in the cave. Abruptly large cracks zigzagged across the eggs. Then rubberlike surfaces stretched and pulled until it popped. Rolling over and over from the activity inside, a small head managed to break the exterior. Large eyes looked around in wonderment. Desperate to be free of the shell, two front legs and shoulders wiggled to freedom. Not realizing he had a tail and two hind legs, the baby dinosaur groaned as he freed himself from the egg. Tired and weak, he lay on his side. Several times he managed to stand but just flopped over.

In the meantime, the other egg was bursting as the second dinosaur struggled for life. She too, winced and groaned, as if being born was painful.

Free from their shells, the babies lay helplessly on the sand. A swishing noise could be heard, like a pendulum swinging back and forth. The cave echoed the sound. Faster and louder the noise became, like a heartbeat. The babies were frightened. Looking down with wide eyes, discovery took place. Two energetic tails were swishing quickly back and forth across the white grains of sand. Encouraged, the babies struggled to their feet. Wobbly and shaky, they slowly regained their strength.

Up on all fours at last, they began to notice their surroundings. The small pond of water beckoned them. Heartily they drank, as their two front feet sunk into the bank.

Outside, the babies could see the sunshine. Tall marshalls of grass grew right up to the entrance of the cave. Still weak and unsure, they broke the automatic curtain to the outside world.

Ravished with hunger, the baby dinosaurs ate the tender blades of grass. Too full to move, they rolled over and went to sleep in the sunshine.

As the afternoon wore on, the temperatures dropped. Waking up cold, the baby dinosaurs struggled to their feet. Instinct drew them back to the cave. Once there, they huddled together. Now safe and warm and contented, the dinosaurs seemed to know that this was their home.

Early morning brought the chirping of birds, the blooms of flowers and the buzzing of bees. Alive with nature, the forest beckoned. Hungry once again, the baby dinosaurs left the cave to eat. Now stronger, tails began to swish like a puppy dog's. Wrinkles disappeared from their yellowish-brown skins. Most of all, they learned their tails would balance them as they went along.

Not knowing danger, the baby dinosaurs fed without a worry in the world. Beneath them a turtle was crawling along. Frightened, they gave out a bellowing sound. Surprised at their own voices, they looked at each other. Lifting their clubby feet, their first instinct was to step on the turtle. Even though the dinosaurs were only as big as a large collie dog, they were ready to defend themselves. Receding inside it's shell, the turtle laid low until the danger had passed.

Each day the dinosaurs would venture farther and farther from the cave. A big world was out there to explore. Soon, they found the large, green meadow with pesky butterflies that tickled their noses. Snakes crawled beneath their feet. Ants built their mounds. And tadpoles swam in the stream. Each day was a new adventure as the dinosaurs continued to grow and thrive.

Reva would be well pleased. Alpha and Omega had faced and conquered their great new world.

42 - Alpha And Omega

Chapter 5

Nestled in a valley, in the foothills of the Rocky Mountains, is a small farming community called Glennview. Each summer the Boy Scouts from this town go on their annual camping trip. Forgotten for a while are the baseball diamond, swimming in the lake, movies and television. And like any small town, whenever something is happening, everyone becomes involved. Buzzing with activity, excitement filled the air as the boys prepared to leave.

Skeeter Davis was no exception. He began throwing things out of his closet right and left. His mother stood in the doorway watching. Shirts and shoes and underclothes were strewn on top of the bed, on the dresser, and on the floor. Mrs. Davis put her hands on her hips, as she always did when she was upset.

"Skeeter Davis, you put that stuff right back!" she demanded.

"But Mom," he pleaded, "I'm looking for my nylon jacket."

Her short, blond hair bounced up and down around her cheeks and her green eyes flashed, as she gave out orders. "Everything you need is in the basement, backpack, sleeping bag and all. Clean up this mess. I've got to go now." She wiped her hands on her apron "I have cookies in the oven."

Skeeter couldn't wait. This was his first trip. Only nine years old, but he looked older. Perhaps being a little chubby made the difference. He cleared the clothes off his bed. Looking in the mirror, the boy could see his flour, curly hair and blue eyes. Yes, he was growing up.

Skeeter wished his father was going along, but he had to work at the sawmill. Only last night, sitting on the front porch, his father had told him all about the trip.

"See that small mountain?" and he pointed toward the Rockies. "Looks as if someone sheared it right in half. We call that mountain Mollies Table. It is flat on top and bigger than a cow pasture. That mesa is where you'll sleep."

Skeeter hung on to every word. He loved to watch his father. He wiggled his dark, brown mustache when he spoke. Of course, his father was never without his knife and a whittling stick.

"Now Henry Garret is going to be your guide," his father told him.

Henry was tall and lanky. His clothes never seemed to fit. They draped around him like a blanket on a bean pole. A man in his fifties, Henry had speckles of gray running through his short, chestnut hair. More or less a

loner, his hazel brown eyes seldom smiled. He did odd jobs around the town. Most of his time, though, was spent caring for his aged mother.

"Why Henry?" Skeeter questioned his father.

"He's a good guide." His father's blue eyes sparkled under his heavy eyebrows. "Did you know Henry guided my first trip to Mollies Table?"

"But Henry, you know Dad, he's sorta, you know, funny and walks with these great big strides."

"Yes, I know son." His father whittled away. "Henry may be a little loose at the seams, but he's a good man, and don't you ever forget it. Incidentally, Henry's hobby is taking pictures, especially ones that deal with strange and unusual rock formations."

"Yeah, I've heard about them too, Dad. He actually took pictures of a rock he said was balanced on one corner." He probably turned the picture upside down and thought people would believe him. Unreal! Skeeter just shook his head.

Smoothing back his dark brown hair, his father continued. "Now getting back to your trip, even though you may think you can't get lost, looks are deceiving. Must be five miles or more to Mollies Table. After you've climbed to the top of the mountain, you'll build a large fire. This way we'll know you arrived safely. Is Doby going?"

"Oh yes, Dad. He wouldn't miss this for the world."

Doby Anderson was Skeeter's best friend. Although much smaller than Skeeter, he was fast and agile. Doby was always the first to be picked for all the games. His straight brown hair literally flew in the breeze, when he made a home run.

"Now pay attention to Mr. Duncan. Skeeter," his father instructed. "He's your Scout Master. If he says stay together, do it. And if I find you haven't obeyed his orders...."

"Yeah, I know Dad, I'll get all my privileges taken away."

Mr. Duncan was quiet and reserved. He was tall and lean with a pointed nose and glasses. Being a pharmacist at the local drug store, Mr. Duncan always had a welcome smile along with the knack of making people feel at ease.

"As you know, Mr. Duncan hasn't had much experience," his father explained, "but I'm sure he'll do a good job. Just cooperate."

"Gotcha, Dad."

Skeeter went upstairs to bed. He couldn't wait to get started on the trip.

Daylight found the scouts and Mr. Duncan and Henry at the town square. Dressed in their heavy shoes, baseball caps, jeans and checkered shirts, the troop was eager to move out. Up went the flag. Backpacks and gear spread across the grass, as they said the Pledge of Allegiance.

"May I have your attention boys?" Mr. Duncan's voice was loud on the bullhorn. "We will line up in single file. Henry will be first, followed by some of you bigger guys. I will stay behind with the smaller ones. You will not rest unless we say so. If some of you feel this trip will be too much for you, back out now. Once we get going, there will be no turning back, unless of course, there is an emergency. Henry and I both have first aid kits. Make sure you have everything. Check your canteens now. Water is a must. No water is available until we reach Mollies Table. Let's be on our way."

Single file, the scouts marched down Main Street. People from the town gathered on both sides to see them off. To keep in step the boys began to sing.

We Are The Scouts Of America

We are the Scouts of America.

We blaze the unknown trails

And backpack in the wilderness;

O'er land and hills and dales.

We are the Scouts of America.

We march and sing and pray.

'Cause we are the Scouts of America,

And we do good deeds each day.

Bridge

We honor the bear and the eagle too;

As we stay to survive.

With hopes and dreams and courage now,

Our banners we hold high.

Chorus

We are the Scouts of America.

Wherever we may roam.

Our hearts are with our countrymen,

And with the folks at home.

We are the Scouts of America.

We march to a distant drum.

We are the Scouts of America.

And our work has just begun.

As the trek advanced, the town was left far behind them. Scrub pines vanished. Taller pines took their place. Hills became steeper. The boys began to tire out. Several times Skeeter almost fell backwards from the weight of his backpack. Thinner air made it harder to breathe as the elevation began to reach the ten thousand foot level. Panting and struggling, they went up one hill and then another.

At long last Mollies Table came into view. Shouts and war whoops filled the air. Although hungry and tired, the troop was happy. They had accomplished their goal. Mr. Duncan once again got on his bullhorn. "Thank you for your cooperation. Without teamwork, we would never have made the trip. You boys have been wonderful. Rest a while, then we'll proceed to the top of Mollies Table. We must get up there before dark. Each one of you will look around for wood and carry some with you on the way up. That is all. We'll continue climbing again in ten minutes."

Climbing was gradual as the trail circled and spiraled upwards. One by one, the scouts reached the top. The sun was ready to drop behind the mountains. Mr. Duncan's timing had been perfect. Just like Skeeter's Dad had told him, Mollies Table was flat and much bigger around than a baseball diamond.

Mr. Duncan instructed the scouts on how to build a fire, as the Indians had done many years ago, long before matches. A pulley was set up on a flat piece of wood causing it to rest on another. When the pulley was moved fast by hand, the wood began to smoke. Soon, the flames began burning the wood. This was placed under the stack of wood the scouts had gathered. A bonfire was started.

At last the scouts could relax. Never had hot dogs and marshmallows tasted so good. Singing and story telling followed. However, some boys couldn't wait. Skeeter and Doby crawled into their sleeping bags and fell asleep before they could even zip them up.

Morning found Skeeter and Doby up early. Anxious to go down and look around, they tried to slip away unnoticed.

Mr. Duncan, was awake, but wasn't dressed yet. Henry was ready to leave, too, with his camera. Mr. Duncan had surprised the boys. He knew they were up.

"As long as you're going down boys," Mr. Duncan said, "gather up the canteens. We'll need the water for breakfast."

"I'll go with them," Henry said. "They might get lost. If they wander into the caves, there's a lot of twists and turns."

After finding the stream that ran toward the meadow, Skeeter and Doby began filling canteens. All of a sudden Skeeter looked up. He gasped. Standing before him were two of the strangest looking animals he had ever seen. Doby was downstream a ways. Skeeter called out to him.

"Doby! Look!" Skeeter was almost speech- less. He called again.

"Doby! Look!" and he too, gasped. "Let's get out of here Skeeter. They may be dan- gerous."

The baby dinosaurs stood and cocked their heads to one side. Large, tantalizing eyes looked at the boys.

"They don't seem to be afraid of us," said Skeeter. "See how their eyes are looking around? It's almost as if they're smiling."

"If I didn't know there wasn't such a thing," Skeeter declared, "I'd say these animals were dinosaurs."

"Yeah," Doby agreed, "tails and all."

Skeeter stood up. "Dare me to touch one, Doby."

"All right, but just watch those tails." Doby warned. "You could be thrown into the next county."

Doby stood there watching the dinosaurs swish their tails back and forth. Unlike a dog, their fanlike motion brushed and spread the dirt. Right now, these animals were really charged up.

Gradually, Skeeter got up enough nerve to touch one of the animals. He stroked the dinosaur down his neck and on his back. Skeeter rubbed the yellowish-brown velvet skin that felt like a chamois. Rolling their large eyes, the dinosaurs seemed to enjoy the attention.

Henry had been watching. He too, was surprised at the strange looking animals.

"Lead them to the tall grass," Henry motioned. "We'll take some pictures."

Not much coaxing was necessary, for the dinosaurs followed the two boys. Henry took the picture.

"Now, we must not tell the other boys." Henry said. "It will cause a lot of confusion and someone could get hurt. Let's keep it our secret for now."

Skeeter and Doby both agreed.

Laden with canteens hanging from their shoulders, the boys went clanking up the trail to Mollies Table.

By now, everyone was awake. Breakfast was started. Hungry once again, Skeeter and Doby could smell the aroma of freshly cooked bacon. Flapjacks swam in maple syrup as the boys savored their early morning meal.

Packing up once more, the long journey home awaited them. Sworn to secrecy, Skeeter and Doby almost burst at the seams wanting to tell someone about the animals they had discovered. A promise was a promise though. Perhaps when the film was developed, the other boys would get a chance to see the animals.

Mr. Duncan stood among the scouts once again. After he captured their attention he began to speak. "As you know, each month we have a lesson on survival. Last month it was first aid. Today will be on how to survive in a forest. A very simple procedure to follow, is your shadow." He straightened his cap and looked toward the east. Mr. Duncan continued, "When this sun comes up in the morning, shadows fall to the west. About noon, you and your shadow are almost one. In the afternoon, shadows fall to the east. If you think you are lost, get in a high place and check the shadows. They will tell you what direction to go. Leave nothing to chance, as in the fairy tale Hansel and Gretyl, you can't follow a path of bread crumbs to safety."

After Mr. Duncan was finished talking, the scouts followed Henry down the trail. Everyone had fun. Next year, they hoped, would be even better.

Chapter 6

The white, two-story frame house on Main Street had never looked so good to Skeeter. He felt like he could sleep for a week. Tired, dirty and hungry, the boys were glad to be home.

Skeeter knew his little sister Sissy would bombard him with questions. He wanted to avoid her at all costs. Like a constant toothache, the small seven-year-old never stopped. She followed him around like a tomboy, spied on him, and tattled too. Her brown ponytail would flip and flop, as she wrinkled up her freckled nose and gave you a look of "I told you so." Sissy loved to argue and never stopped badgering Skeeter about one thing or another.

Mama had found the family seated at the breakfast table. All except Mr. Davis who had gone to work. Mrs. Davis was anxious to hear about the trip to Mollies Table.

Skeeter wanted to talk to his mother, alone, in the worst way. Sissy, again, oh well, he thought, I guess, there can't be any harm. She'll find out anyway.

Skeeter began to tell his mother about the strange animals he and Doby had discovered.

"They were a yellowish-brown, with long tails that swished the ground." Skeeter's eyes were wide with enthusiasm, as he told about the dinosaurs. "Doby dared me to rub one. I was scared, but the animal stood still as I stroked his back. They looked like dinosaurs Mom, honest they did."

Sissy chirped in, "Ever since you went on that field trip to the museum with the school, it's dinosaurs, dinosaurs, dinosaurs."

"That's enough, Sissy," her mother said. "Stop that bickering, right now!"

Skeeter continued, "Henry wouldn't let us tell the other scouts. He was afraid someone might get hurt."

"I can understand," his mom said, as she sipped her coffee. "All that many scouts are hard to keep track of."

"Anyway Mom, Henry took some pictures of Doby and me and the animals."

Skeeter looked at Sissy. "You'll see, Sissy, when the film is developed."

"Yeah, I'll bet," she taunted.

Skeeter, my son, listen very closely to what I have to say," his mother told him. "I'm not going to call you a liar, but you'll have to admit your story is pretty farfetched. I'm sure, somewhere, there is a logical explanation. Until then, consider the subject closed."

Every day the boys met Henry at the drugstore to see if the pictures were back. Finally, the day arrived. Henry held the packet of pictures in his hands. The boys jumped on one foot and then the other.

"Go through them Henry," Skeeter blurted out. "We can't wait to see the animals."

Shuffling through the pictures, a puzzled look crept onto Henry's face. Doby just stood there wondering what was going on.

"They didn't come out," Henry said.

"Oh no!" cried Skeeter. "They just have to."

"Are you sure?" Doby asked.

"Yes," Henry said. "Look for yourself." Skeeter shuffled through the pictures.

"He's right, Doby. Look at this one by the tall grass. All you can see is you and me and grass and the animal's eyes."

"Where are the animal's eyes?" Doby asked.

"See? They look like the center of the flowers."

Confused and unhappy, the boys walked down the street.

"What am I gonna tell Mom and Sissy and Dad?" Skeeter asked.

"I'm in the same fix," Doby said, "but there's gotta be a way to prove it to them."

"Henry," Lights went on in Skeeter's eyes. "He'll tell them we're not lying."

"Naw," Doby said. "They wouldn't believe Henry either."

"We've just gotta go back to Mollies Table," Skeeter decided.

"Yes, I was thinking the same thing," agreed Doby. "I want to make sure this wasn't just another bad dream."

"Aren't you afraid of getting lost?" Skeeter asked.

Doby cocked his head to one side and looked at Skeeter with his large brown eyes. "I think we can find our way over all right. It's just having an excuse for leaving. If not, we'll have to sneak away."

"Maybe we can take some horses too," suggested Skeeter.

"All right," Doby agreed, "now you're cookin'." The boys slapped their hands together. "We'll think up an excuse and plan to go Saturday morning."

Friday night came and the boys still hadn't figured out a solution. Even though the boys knew they would be punished, their bound to's were worse than their can't help it. Sneaking away would have to be the answer.

Early in the morning Skeeter crawled through his window. Still dark outside, he slowly and carefully climbed down his mother's rose trellis, from his upstairs room. All the food he was able to get was a pocketful of cookies, and any minute now they might be crumbs. He sure hoped Doby brought some food.

Rushing to the barn, Skeeter wanted to saddle his horse as fast as possible. He didn't want to turn the light on, but he couldn't see. Working feverishly, everything went well, until the time came to cinch the saddle. His father had always done that. Skeeter found he didn't have the strength.

Skeeter heard a noise. His heart flew to his mouth. He just couldn't get caught now. The barn door opened. It was Doby. Skeeter breathed a sigh of relief.

"Oh, it's you Doby."

"Turn off the light," Doby said, "Let's get out of here. We'll ride bareback."

Leading the horse down behind the pasture, the boys had to climb a fence to get on and ride. After they had ridden a couple of blocks from home, they whipped the horse with the reins. Galloping down the street, the boys headed toward the mountains.

The sun had been up a couple of hours before the boys' disappearance was discovered.

Sissy woke up early to use the bathroom. She saw a light on in the barn. Checking up on Skeeter, she ran to his room. Skeeter wasn't there.

"Wake up Dad! Dad wake up!" Sissy shook

her father. "Skeeter's gone!"

"He's what?" her father asked.

"He's not in his room!"

Putting on his clothes, Mr. Davis went to check on the boy. No, he wasn't in his bed.

"I'm going to call Doby's house," he told Sissy, "and see if he spent the night."

"Ebbie, get up!" Mr. Davis called to his wife. "Skeeter's gone!"

Rubbing her sleepy eyes, it took a few moments for Mrs. Davis to realize what was going on.

"The Anderson house," he told his wife. "Doby is missing, too."

"Where in the world do you suppose they went?" she asked her husband.

Rolling her eyes and looking smug, Sissy blurted out. "They probably went to find those animals they thought they saw at Mollies Table. The pictures didn't turn out, you know."

"Now Sissy," her mother broke in, "we can't be sure where they went."

"I only know one thing," said Mr. Davis, "we've got to find the boys. Go look in the barn Sissy, and see if Bessie is there."

Still in her pajamas, Sissy ran to the barn. When she came back, the youngster reported to her father. "The horse is gone, Dad, and the saddle is on the floor."

"Thank you, Sissy," her father said, "I'll go down to Henry's and see if he knows anything about the boys."

Henry lived in a small brown cottage, with flowers blooming in the front yard. Mr. Davis knocked hard on the door. Only partly dressed, wearing jeans, a T-shirt and suspenders, Henry answered.

"Good morning, Cliff. What brings you to my house so early in the day?"

The two men didn't bother to go inside, but stood in the open doorway to talk.

"Skeeter and Doby are missing," Mr. Davis told him. "They've taken Bessie, our horse. Do you suppose they went back to Mollies Table?"

"I don't know," Henry answered. "I know they were pretty upset last week when my pictures didn't turn out."

"Yes, the boys did come home with a fantastic story about two strange animals."

"Yes, there were two animals that I'd never seen before," Henry told him. "Then again, I don't know much about wild animals. I wouldn't even recognize a cougar if I saw one."

"Could I ask you a big favor, Henry?"

"Sure, Cliff."

"Would you take your horse and ride to Mollies Table and look for the boys?"

"Yes," Henry replied, "but I'll have to get someone to stay with my mother. She's been ailing."

"I'll send my wife Ebbie down," Mr. Davis told him, then I'll search the entire town."

"I'll leave right away," Henry promised.

In the meantime, Skeeter and Doby had reached Mollies Table. Not seeing the dinosaurs, they looked around. Directly across from Mollies Table was a cave. Going inside, they found a small pond. Next to the wall were two empty gray missile-shaped containers. Next to them were two large grayish looking shells that had been cracked and broken. Skeeter spotted them first.

"Doby come here and look at this!" Skeeter was excited. "Where do you suppose these shells came from, and these funny looking missile things?"

"I don't know, Skeeter, but there's some kind of writing on the shells. See this one? It spells out A L P H A." Moving to the other shell, Doby spelled out O M E G A. "Look Skeeter, there's something else written on them too, but I can't make it out. I've never seen writing like this before."

"Maybe they're from outer space," Skeeter said. "The animals could have hatched from the eggs. I've never seen shells this big. They're a hundred times bigger than a chicken's. I'll bet they brought the eggs in those missile shaped cages."

"Yeah, I'll bet that's what happened, all right," Doby agreed, as he nodded his head up and down.

Gathering leaves and branches, the boys covered the containers and the egg shells. They walked out of the cave and sat on a large rock.

"I'm hungry," said Skeeter, "and all I've got is some cookies."

"I don't have a chance to get much either," Doby said, "but here's two pieces of chicken."

Chicken and cookies sure tasted good.

Looking up, the boys saw the Dinosaurs coming from the meadow. After crawling up a hill, Skeeter and Doby watched the animals. Walking clumsily along, the dinosaurs went into the cave.

"That cave must be their home. It's a good thing we got out of there when we did, Doby. Even though the animals seem tame enough, they could turn on us. Anyway, we know now, this wasn't just a dream. What do you say, Doby? Does it feel better start for home?"

Leading the horse up to a small hill, the boys climbed up until they could get on Bessie's back. Off they went, downing one slope after another. Riding around for at least an hour, the boys discovered they were in the same place as where they had started.

"We've been going around in circles," said Doby. "We're lost!"

"Oh no!" cried Skeeter. "What are we going to do?" Big round tears rolled down his chubby cheeks. Never in his life, had he been so scared.

"Didn't you hear Mr. Duncan's lesson on survival?"

"No, Doby," and Skeeter was all choked up. "I was thinking about the animals, not survival."

"I didn't hear him either, Skeeter. We're really in a jam."

Just then the boys heard the heavy breathing of another horse. They turned and looked. It was Henry riding toward them. Never had the boys been so happy to see a human being. Screaming and waving their hands, the boys hollered out. "We're over here Henry! We're over here!"

"I've come to get you boys. Your families are worried." Henry dismounted. "Let me rest for a while and eat lunch, then we'll go home. In the meantime, I want to talk to you guys. Get down off your horse. I've brought some sandwiches. I thought you might be hungry."

Sitting down on the same big rock where they had ate before, the boys listened to what Henry had to say.

"I think I've figured out what happened to the pictures." Henry said handing the boys a sandwich. "What color were the animals when you saw them Skeeter?"

"They were a yellowish brown."

"What color were the animals, Doby, when you saw them?"

"They were gray, of course."

"When I saw them," Henry said, "they were a reddish brown. Have you boys ever heard of a chameleon?"

"No, I haven't," said Skeeter. "What about you, Doby?"

"No, I haven't either."

"It's a lizard that changes to the colors of his surroundings."

"Oh," said Skeeter, "when I saw the animals," and he pointed toward the east, "they were in front of that yellowish-brown rock."

"Yeah," Doby said, "when I saw them, they were standing by the gray mountains."

"When I saw them," said Henry, "they were standing by a reddish-brown hill."

"Really, it dawned on Skeeter. "Now I understand. When you took the pictures in the tall grass, the animals turned green. That's the reason they didn't show up in the picture."

"Right!" Henry said. "Now that we've had our lunch, let's go home."

Skeeter was sorry about how he misjudged Henry. His father was right. Henry was a good man.

Chapter 7

Skeeter might as well have been grounded for life. He wasn't allowed to see, or even speak to Doby. Even the windows in his room had been nailed shut. All of his privileges were taken away, including television. Skeeter's father blamed all his wild stories on the tube. On top of that, his father didn't believe anything the boy had told him about the animals.

Lashing out at Skeeter when he came home, his father was very angry. "First, you come home from your scout trip and tell us a fantastic story about seeing two baby dinosaurs." Pacing back and forth across the floor, his father's anger grew. "Then, you not only sneak out of the house, but have the nerve to take the horse. Now you expect us to believe yet another cockamamie story. They came from outer space. Perhaps my grandfather was right... spare the rod and spoil the child. In good stint in the woodshed, young man, would do you good. Now go to bed. I don't want to see or hear from you, until you can learn to appreciate a good family and respect my home."

School would be starting soon. A new year would begin, as Skeeter entered the fifth grade with a different teacher. Although busy thinking about school, thoughts of the animals still lingered in his mind. Would they freeze to death in the winter's snow? How would they eat? What would happen to the animals? All these questions and no one to answer them, not even Doby.

* * * * *

As the summer turned into fall, autumn brought with it an array of colors. Leaves turned from green to gold, then from gold to brown. Falling gently, one by one, they covered the ground like a brown paper quilt. Alpha and Omega began to feel the drop in temperatures. Only a few hours to feed and then back to their cave. The animals were getting ready for hibernation. Growing fast, over the summer the dinosaurs had reached a height of about 12 feet (3.6 meters) and a length of 14 feet (4.26 meters) or more. Being large, many of the wild animals bothered the dinosaurs. The deer and the coyote just stood and watched, but kept their distance. The cougar, however, was another story. The tawny, brown cat would spread across a ledge and growl at the dinosaurs. He seemed to know they didn't belong in the mountains.

One day, almost in defiance, the cougar lunged at the animals. Ready to fight, the cat was hoping to land on the dinosaur's back. He landed all right, but bounced off and landed hard on the ground. Afraid for the first time, the cougar lay petrified. Lifting his large front foot, the dinosaur rolled the cat around like a piece of bread dough. To the dinosaur, this was like playing with a kitten. When finally released, the cat lay helpless on the ground. Gaining strength once more, the cougar managed to get on his feet. He looked up at the dinosaur and growled.

Abruptly the dinosaur turned, and with a flick of his giant tail, the cougar was thrown through the air and landed on a split bank. The cat was knocked out of the cat. After that there was too, kept his distance. The dinosaur looked around as if to say, "Where has my playmate gone?"

When fall turned to winter, the dinosaurs had already crawled inside their cave and gone to sleep. Trees and grass and meadows had gone to sleep in Glennview, as well. Snow began to fall, and the long winter kept the town in a veritable deep-freeze.

Skeeter and Doby were slowly gaining back their privileges. Now they enjoyed sledding, tobogganing and ice skating on the lake. Once in a while they even watched television. What was more important, Skeeter was on speaking terms with his father.

Gradually, the long winter turned to spring. Along with springtime came spring fever. The boys were itching to play baseball and swim and fish and, of course, go back to Mollies Table. Even though Skeeter and Doby couldn't speak out loud about the dinosaurs, in the back of their minds, they were still planning and scheming. Looking forward to going on the scout's annual trek, Skeeter and Doby wanted to show Mr. Duncan and the boys the strange animals. If the animals weren't there anymore, they could at least show them the eggshells and the capsules.

Eagle Scouts were graduating. Everyone in town turned out for the event. After the presentation of the medals and the certificates, Mr. Duncan gave a speech.

"We are very proud of our boys. Very few achieve this great honor. In the years to come, we hope these young scouts follow in their footsteps. I am sorry to announce we are not going to Mollies Table this year. However, we are going to a Jamboree at Round Mountain. Scouts from all over the state will be there. I'm sure you boys will have just as good a time, if not better."

Skeeter and Doby's hopes vanished. They wanted desperately to prove themselves to the town and their parents. Everything seemed to be working against them.

Eventually the last day of school arrived. Skeeter knew his mother and father were planning a vacation, but he didn't know when. He did know they wanted to plant their vegetable garden before the family left. If he and Doby were going back to the mountains, they must make their plans. This time, they couldn't afford any

foul-ups.

Skeeter and Doby took long walks and talked by themselves. Finally, Skeeter came up with an idea. Sitting by the lake and watching the ducks, he told Doby of his plan. "We are going fishing, Doby, only we aren't going fishing."

"How's that Skeeter?"

"We'll tell our parents we're going fishing. We'll take the small rowboat, row down as far and then go to Mollies Table."

"Hey Skeeter, you're all right."

The boys took off their shoes and tangled their feet in the water.

"But, we won't be bringing home any fish," Doby said.

"Lots of times people fish and don't catch anything." Skeeter wiggled his toes in the water. "We'll just say the fish weren't biting."

"Boy are you smart, Skeeter."

"I'll ask my Dad tonight," Skeeter said. "I

just hope we can pull it off."

"Are you sure, Skeeter, if we go, we can find our way back? I don't want to get lost again. I'm prepared this time, Doby. I have the Boy Scout manual."

"We're all set then, Skeeter!" They slapped their hands together in agreement.

After the evening meal was over, Skeeter asked to talk to his father. Mr. Davis had already settled back in his easy chair to watch the news on television.

"Yes, Skeeter, but make it snappy."

Skeeter was nervous. He wiggled around on the ottoman at his father's feet. "Doby and I want to go fishing and we want to take out our small rowboat."

"Oh," and his father looked at him and thought a minute. "Well, I don't know Skeeter, that rowboat has been dry-docked since last summer. You'll have to test it for leaks. I have sealer and whatever you need for repairs in the barn, but couldn't you just fish off the bank?" his father asked.

"I think we'd do better at the cove, Dad. Anyway, I don't want Sissy hanging around. You know what a pest she is."

"All right then," and he pushed his recliner back, "if you're willing to fool around with that boat, then you can go."

"Thanks Dad," Skeeter said. "Doby and I will start working on the boat tomorrow."

Dragging the boat out of the barn took quite an effort, however, they managed to drag it to the back yard. Filling it up with water from the garden hose, the boys let it stand a while to see if the water level would sink. Sure enough, the water began seeping out a little at a time.

"We've got some leaks," Skeeter said. "Might as well dump out this water and start repairing the darn thing."

"All right, Skeeter," Doby said, "just hope you know what you're doing."

"Dad told me how to repair it, Doby. Plug up the holes and put sealer on it."

Sanding and scraping, the boys worked most of the day to get the boat in order.

"We'll let it dry overnight, Doby," Skeeter said. "In the morning we'll fill it with water again and test for leaks. If all goes well, we can leave day after tomorrow."

Carrying the boat upside down above their heads, the boys walked down the street toward the lake. Late May was still quite cool, so Skeeter and Doby wore their jeans, flannel shirts and baseball hats. The boys had taken their fishing poles and oars and their lunch to the lake before taking the boat. Elated, Skeeter and Doby were all set to go.

"I'm a little scared, Doby."

"Why Skeeter?"

"Before I left, my father talked to me."

"What did he say, Skeeter?"

"He said: 'Now, I'm going to put my trust in you two boys. No more hanky panky to the mountains. If you pull a stunt, as you did last year, I might even be the juvenile home for both of you. Please, Skeeter,' he begged 'Don't let me down. Go now and have a good time'."

"That does give you a funny feeling in your stomach," Doby admitted. "I'm almost sorry we planned this trip."

"But Doby, we can't back out now."

Slowly, the boys slid the boat into the water. Paddling not too far from shore, they were on their way. Soon they came to the cove. Stashing the boat and the fishing gear, they headed toward the mountains.

Canteens were filled with water and their Moms had packed nice, big lunches. Full table was in sight. Thankfully they had made it. The boys sat down to rest.

Hearing a noise Skeeter and Doby looked up. Standing before them were two giant animals. Scared because the dinosaurs had grown so large, they quickly scampered up a hill. Shaking their large heads back and forth, the animals bellowed at them. The word alone nearly drove the boys crazy.

"Now what are we gonna do, Skeeter?"

"I don't know Doby." Skeeter's voice was shaky. "We better head for home. I'm so tired though, I don't know if we can make it."

"Me too," said Doby. "Besides, I'm hungry."

Even though the boys had crawled to the top of the hill, the dinosaurs kept coming. Trying to get away, the boys went down the other side.

"Maybe they won't see us Doby. We'll sit here for a while and wait until they go away."

Wishful thinking didn't make it so, for the animals kept coming. Evidently, the dinosaurs hadn't forgotten the boys. Hiding behind a large boulder, Skeeter and Doby figured they couldn't be found.

"What are they doing now, Doby?" Skeeter asked.

Doby peeked out. "They're feeding on some grass on a hill."

Standing on their two hind legs, the dinosaurs could rear to browse. Using their front legs as arms, the animals then propped themselves up.

"What do we do now, Doby? We can't stay here all day."

"Run like hell!"

Off the boys went. Skeeter's chubby little chest and hips bounced up and down. His face was red. His eyes were full of tears and he'd lost his cap. After running for a while, the boys looked back. The dinosaurs were still following them. Scared and desperate, the boys never stopped running.

Chapter 8

Jake and Abagail Carlin lived on the first small farm at the edge of town. An elderly couple, they had spent their entire lives in the town of Glennview.

Mr. Carlin was sickly and couldn't get around too well. He was bent over and thin. He walked with a cane and his face was wrinkled and creased like an old tire track. Eyes sunk back in his head, and hardly able to get around, Mr. Carlin could barely manage to do his chores. Mrs. Carlin, on the other hand, was short and plump. Being kind and friendly, her blue eyes always smiled. Mrs. Carlin wore her silver-gray hair neatly in a bouffant, with a French roll and dainty combs which were pinned close to her head.

At last, Skeeter and Doby and the dinosaurs arrived in Glennview. Trying to hide from the animals, the boys headed for Carlin's barn. They ran inside and shut the door. Sitting very still, but breathing hard, they peeked through the cracks of the unpainted barn. Yes, the dinosaurs were still coming.

"I heard somewhere, Skeeter, if you go through water, an animal can't pick up your scent, at least hound dogs can't. So, what do you say, should we make a run for the lake?"

"Good idea, Doby. Now let's get out of here."

The boys no sooner left the barn, than the dinosaurs came plunging in. Both sides of the building collapsed. Then the roof came crashing down like fall timber. Dust and silly billowed everywhere. Bewildered, the cow took one look and high-tailed it back to the pasture.

Running as fast as they could, the boys thought if they climbed the fence to the pig pen they could make better time.

"Oh my gosh, Doby! I've got poop clear up to my ankles."

"Me too, Skeeter, but we'll gotta keep going."

The big floppy animals walked right through the pig pen and broke all the fences. Little pigs, the boar and the mother pig ran squealing in all directions.

The boys almost got trapped in the chicken coop. Followed by the dinosaurs, Skeeter and Doby barely made it out of the gate before the animals came lumbering in. Chicken feathers flew through the air, like someone having a pillow fight. Trying to escape, but not able to fly over the coop, the chickens hung on desperately to the wire mesh fence, squawking and flapping their wings wildly.

Running toward the water tower on their way to the street, the boys scooted under the trusses. Still wobbling along and swishing their giant tails, the animals couldn't walk under the water tower, but knocked it over. Water, as well as pigs and chickens and two frightened boys, followed by two dinosaurs, made a strange looking parade down the main street of Glennview. Bellowing and braying, the huge grayish-brown dinosaurs alerted the entire town. Some of the people came out to watch, while others went into their houses and looked the doors. After seeing animals as large as a two-story building, no one could blame them.

Finally, Skeeter and Doby arrived at the lake. Swimming hard, they reached the wooden dock. Holding on to the beams that supported the wharf, the boys were trapped until some kind of help arrived.

Stopping at the edge of the water, the dinosaurs sank their feet in the mud and began to drink. Standing and waiting, they swung their big heads around to look for Skeeter and Doby.

Overhead, the boys heard a spinning, whirring sound. Looking up, they saw the bottom of a giant space ship. The craft was so large that Skeeter and Doby thought the sky was falling in on them.

Hovering in mid air, a large ramp was slowly lowering to the ground. The boys couldn't see what was happening, but heard a sweet, wonderful voice.

"You can get out of the water now boys. The dinosaurs have been put under a hypnotic beam. They cannot move."

By now the town's people had begun to gather by the lake. Slowly, the boys swam to shore. Dripping wet and shivering, someone handed them blankets.

Standing on the ramp was one of the most beautiful women Skeeter and Doby had ever seen. She was dressed in gold tights, with a gold tunic and a gold belt. She had on a quaint little Peter Pan collar. Flowing like brown silk, her hair shined and glimmered in the afternoon sun. Standing behind her were two men. They were dressed in dark military looking uniforms. Reva began to speak.

"My voice will sound slow. While I am talking to you, my language is being translated into English. She had the transmitter in her hand. "My name is Reva. I came from a planet called Plateo. A long journey through space. We have traveled many months to reach the Earth. Behind me on the right is Gareth. He is the commander of our ship. On the left is Ammone, he is the leader of our space squadron. We have come to take home the dinosaurs. I am the daughter of the President of the Federated Zones of Plateo. My father's picture is projected in the air above you."

Everyone looked up. President Karabe's face was floating in the air but there was no screen. Reva continued, "On our planet we had a nuclear war. All of our dinosaurs were destroyed. We found eggs that had not been contaminated and brought them to Earth for incubation. It was not our intention for anyone to discover the animals. However, two of your boys discovered them by accident. Evidently, they petted the dinosaurs when they were babies. Dinosaurs do not forget. Seeing the boys again, they followed them home. We went to the cave first. Not finding the animals, we knew they must be here, for this is the closest town. We are sorry for the inconvenience and will compensate you for all damages."

Standing with their mouths open, the town's people were speechless.

Reva beckoned Skeeter and Doby to come closer. "And what are your names?" Reva asked.

"I'm Skeeter Davis and this is my friend Doby Anderson."

"You boys had quite an experience," Reva remarked.

"Yes, we did," said Doby. "We ran for our lives."

"You are two daring young men. I commend you on your bravery and courage."

Skeeter looked around. His mother and father and Sissy were standing there listening. Mr. and Mrs. Anderson stood beside them.

Mr. Anderson was a large, brawny man with light brown hair and blue eyes. He had been a football player in college but was now the coach at the local high school. Mrs. Anderson was petite with large, expressive brown eyes. Dark wavy hair curled around her doll-like face. Still in her apron, T-shirt and slacks, Mrs. Anderson had been preparing dinner when all the commotion started.

Reva began talking once more. "Sliding down the ramp, is a chest full of gold coins. These will pay for repairs to your town, plus build an island for the dinosaurs we are going to you. I have the plans. Skeeter," she asked, "will you come here? Bring your father."

Mr. Davis smiled and stepped forward. Reva gave him the blueprints for the island.

Reva continued, "You will be protected by the animals being on an island. You see, dinosaurs do not swim. In turn, the animals will be protected from the outside world. You will then have the only Plateosauruss on Earth."

Six men in green jump suits came out of the ship to lift the chest and set it on the grass for all the people to see.

Out of the crowd came a short, pudgy man. "What is your name, sir?" asked Reva.

"I am Mayor Stokey."

Mayor Stokey took the long cigar out of his mouth. His hazel eyes were glowing with curiosity below his bushy eyebrows. He took out a handkerchief and wiped the perspiration off his balding head. Pulling his black pants up over his round tummy, the mayor let out a chuckle where he had let them out notches on his belt. The Mayor rolled up his sleeves, nearly splitting his sleeves. He couldn't wait to see the gold coins. Rubbing his fingers over the intricately carved wooden chest, Mayor Stokey noticed the delicately designed crest of the Plateo Empire. It was an insignia of a dinosaur.

"As you can see," Reva continued, "our crest is the dinosaur. We had the only living species known to any planet. Thanks to you people, we will have them again. In all the galaxies the dinosaurs have become extinct for millions of years. Plateosauruses are vegetarians. Being harmless, except for their size and dangerous tails, they are wonderful animals."

Mr. and Mrs. Carlin came to see what was going on. He wondered how his livestock had disappeared and who or what wrecked his barn. Hobbling along with his cane, the poor man barely made it to the lake. Standing beside him in a flowered gingham dress, was his wife Abagail. She had brought her dog Lizzie. The animal was going to have pups and Mrs. Carlin was afraid the excitement would cause her to lose the litter!

Mr. Carlin looked up at the dinosaurs, which were standing lifeless at the edge of the lake. Without warning he collapsed onto the grass. Reva saw what was happening. She signaled for the medics.

"Not my friends, do not be afraid," she told the people, "we are farther advanced in medicine than the planet Earth. Please permit us to take care of this sick man."

"Go ahead," Someone shouted.

Quickly the medics put Mr. Carlin on a stretcher and they disappeared into the space ship. In the medical compound area, he was suspended below computer monitors. Mr. Carlin was placed in the first tube. His vital signs were taken. Moving into the next tube, the diagnostic monitor registered the cause of his condition. Anesthetic was automatically given. Hundreds of laser beams worked together, opening and closing his chest area. Mr. Carlin was undergoing open-heart surgery. Still moving along through the tubes, Mr. Carlin went under the healing beams. Within half an hour the man was placed on a recovery table.

Rising on his own, he stood up. "How do I get out of here?" he asked. One of the medics led him down to the ramp.

Throwing his hands in the air, Mr. Carlin called out to his wife. "Abagail! Abagail! I'm well! I'm well! I can walk without my cane. I feel wonderful!"

He hugged and kissed his wife. He then looked up at Reva. "I don't know what you have done, but you saved my life. I will be eternally grateful."

"The pleasure belongs to us," Reva smiled. "Bring the dog, Mrs. Carlin. I'd like to take a closer look. She's the most beautiful dog I have ever seen."

Reva looked at the brownish collie, with the white and yellow and gold hair running like streaks of silk on the animal's neck and back. Reva knew this animal had been loved and well taken care of.

"Would you like to have my Lizzie?" Mrs. Carlin asked.

"Oh, I couldn't take your pet away from you," Reva said.

"But I want you to have her," Mrs. Carlin insisted. "She's going to have puppies, you know."

"How many will she have?" Reva questioned the lady.

"I won't know until they are born, perhaps eight or nine."

"Thank you, Mrs. Carlin. How could I refuse such a beautiful dog and such an unselfish gift? Lizzie will be put to sleep like the dinosaurs until we reach Plateo. Her pups will be born on our planet."

A hypnotic beam was thrown on the dog. She walked up the ramp of the space ship and into Reva's heart.

Chapter 9

One by one the town's people would disappear from the crowd, then return. In front of the ramp they placed baskets of apples and nuts, jars of dried fruit, jars of home canned vegetables and fruits, along with preserves and jellies were put together by the people of Glennview. Reva looked down from the top of the ramp, and seeing all the bounty, tears filled her eyes.

"Mr. Davis approached her. "We have gathered a few gifts. Would you accept this food as a token of our friendship?"

"Oh, Mr. Davis," Reva said, "How thoughtful."

"Since our gardens have not yet been cultivated," Mr. Davis explained, "and the blossoms are just coming out on the trees, our fresh crops will not be ready to harvest for a couple of months."

"Do not apologize," she told him. "You have been very generous. Thank you one and all." Reva raised her hand. "We shall savor this wonderful food on our long journey home."

Finally, Mayor Stokey opened the long wooden chest. Rays of sunlight reflected on the gold coins and made them sparkle like hundreds of twinkling stars. Reva looked with indignation at the little man who was on his knees, greedily biting and counting the coins. Evidently, he had down on one, Reva finished. She did not have time to dwell on trivial matters, however. Time was of the essence. Once again she addressed the people of Glennview.

"Our time here is very short. If your government learns of our presence, we could be delayed, not for just a day or a week, but perhaps months. At this time we are not prepared to confront them."

"Gareth took me on a tour of your planet. So magnificent and perfect, as your world. You must do everything to preserve it. Eliminate the pollution of your cities and waterways. Try to get along with other countries. Love one another, no matter how different you may be. Do not let your planet be plunged into chaos as ours has been. Hold sacred your ideals and principles.

"Before we brought the dinosaur eggs to Earth, our enemies, the Paradons, did not want to rebuild our planet. Instead, they chose to invade the Earth. We fought a battle in space and won. My father would not let the Paradons take over and destroy your world, as they have ours."

Everyone clapped and cheered.

"Now we must leave you," Reva told them. "In conclusion, we have left our legacy. Until we return, may you walk in the footsteps of the Master."

Reva turned, followed by Gareth and Ammone and disappeared inside the ship. Before they left, even wider and higher doors opened so the dinosaurs could enter. Beams were aimed at Alpha and Omega. Slowly, the animals came to life. They walked up the ramp and entered the craft.

Up until now, Sissy had remained quiet. Rather awe struck at the day's events, she had stayed close to her father and mother. When the space ship started to rise, Sissy began to talk.

"Look Dad, the space ship is rising like a heavenly star. Now it's spinning." Her arm went around in a circular motion as she pointed toward the ship. "Now it's spinning even faster, and going round and round and round like cotton candy. See the hundreds of rainbows dancing along?"

"Yes, my dear," her father said, "isn't it spectacular?"

Everyone watched and waved until the space ship had vanished into the evening sky. People started going home. Mr. Davis looked up. There stood Henry with his camera hanging on his shoulder.

"I haven't seen you all day, Henry."

"I've been taking pictures, Cliff. I'm going to send them to Washington D.C. I know they'll believe our story when they see the photographs."

"Perhaps they will, Henry," Mr. Davis said. "You should have brought your mother. The space people might have cured her illness."

"Oh, but Cliff, she's feeling much better today."

Henry then wandered off to talk to someone else.

Skeeter just couldn't help it. He rolled his eyes and shook his head back and forth. He tried to get his father's attention. "Slow down a minute Dad. Don't walk so fast. I want to ask you a question."

"Yes, what is it Son?"

"Just why is Mayor Stokey biting those coins?"

They both looked at the Mayor, who was still on his knees and oblivious to the world around him. Skeeter and his father both laughed.

"You see Son, when you bite on a gold coin, it makes an indentation on the coin. Of course, if they only look like gold, there won't be any indentation. He's biting them to make sure they're real. Eventually though, they will have to be melted down into gold bars before we can receive money to rebuild our town and construct an island."

"Do you think he'll be there all night Dad?"

"I have no idea," his father told him.

Trying to catch up with Doby and his family, the Davises left the Mayor still on his knees biting and counting.

"Hey! Cindy and Mark!" Mrs. Davis called out, "may we walk with you? I want to ask you something."

"Oh sure," Mrs. Anderson said.

"What is it, Ebbie?" "Why don't our two families go to Mollies Table on a picnic? I've never been to the mountains. Skeeter and Doby can show us where the cave is located and the dinosaur shells and the capsules."

"That's a good idea. I haven't been to the mountains either." Cindy looked at her husband. "How about Mark? It sounds like fun."

Everyone walked and talked until the plans had been made.

Lying in the bed that night, Mr. Davis tossed and turned and wrestled with his thoughts. He had warned Skeeter about going to the mountains and pursuing the nonsense of the dinosaurs. In view of what had happened, he wondered if the boy should be reprimanded. He asked himself some questions. Could he remember back when he was a boy and his father didn't believe him? Yes, there were many times, he recalled. In fact, when he'd found a dollar bill that had blown onto a bush, his father didn't exactly accuse him of stealing, but he had again, he didn't believe him either. Skeeter had taken a great risk, just trying to prove to his parents and the town that he wasn't lying. Feeling the matter should be closed, Mr. Davis went to sleep.

* * * * *

Both families were up, bright and early in anticipation of their trip. Somehow, they had all managed to beg or borrow a horse. Everyone was dressed in jeans and trekking apparel. With caps and hats and cameras and lots of food, they set out for Mollies Table. A bright, clear day promised to make the picnic an enjoyable occasion. Singing and laughing and telling jokes made the trip a pleasant one.

At journey's end, they let the horses drink and graze and everyone went to look for the cave.

"Now, let me show you," Skeeter said, when he started to enter the cave, "where the shells are."

"We covered them with leaves and sticks," said Doby.

"Wow!" Skeeter hollered out. He rubbed his head as if something had hit him. "I can't get into that cave."

"What do you mean?" his father asked.

"Feel and see, Dad."

"Well, Son. There's some kind of invisible curtain over the entrance. Come here, Mark," Mr. Davis said.

Mr. Anderson ran his hand over the clear surface. "You're absolutely right, Cliff, the space people have sealed the cave."

Putting both hands on the sides of his face, Mr. Davis tried to peek inside.

"I can't make out much of anything," he told his wife. "Bring me the binoculars."

Mr. Davis looked through the binoculars. He not only saw the large eggshells and the capsules, but sticking out of the sand were two large eggs. Written on the top of one egg was ALPHA 11. Underneath that was writing also, but he figured it must be in their language as he couldn't decipher the script. Written on the other egg was OMEGA 11.

"Here everyone," he called, "take a look for yourselves."

The binoculars were passed from one to another as each one looked at the new eggs. Skeeter's mother stood deep in thought. "I think, I know what's happened."

"Yeah, Ebbie, I'm sure you do," Mr. Davis commented.

"The dinosaurs laid their eggs. That's what happened. Remember at the end of her speech, Cliff, when Reva was talking? She said something I didn't quite understand."

"What was that Ebbie?"

"We have left our legacy.' Don't you see? The eggs are their legacy."

The men went off with Sissy tagging behind, to talk about building the island.

"We're getting hungry Ebbie," Mrs. Anderson said, "I'll take Doby and we'll start getting the food ready." "All right Cindy, but I want to climb to Mollies Table." Mrs. Davis said. She then, looked down at her son. "Will you come with me, Skeeter?"

"Sure Mom. I'll show you the way."

Going up and around the spiral trail, they both huffed and puffed. Skeeter's Mom was a little chubby, too. Reaching the top, Mrs. Davis stopped to rest a few minutes and then rose to her feet.

"Oh Skeeter, look at this breathtaking view. Have you ever beheld anything so majestic? Mrs. Davis put her arm around Skeeter. "Reva was right. Our Earth is too precious to be destroyed. Take care of it Son."

"I will Mom. Now, can I ask you a question? When the dinosaurs are hatched and on their own island, do you think the world will believe us?"

"I don't know Skeeter," and she looked up toward the heavens. "I only know one thing."

"What's that Mom?"

"We are not alone."

****The End****

The sequel is coming.

Alpha & Omega II

The truth is out there…

www.ingramcontent.com/pod-product-compliance
Lightning Source LLC
LaVergne TN
LVHW011518200926
848484LV00001B/66